HEINEMANN MATHEMATICS 5

Textbook

These are the different types of pages and symbols used in this book and associated workbooks.

1
Using + and − facts to 15

These pages relate to particular groups of workbook pages, developing the mathematical skills, concepts and facts introduced in the workbook. In many places they also extend the theme presented in the workbook.

4
Other activity: addition to 10

These pages provide self-contained activities which need not necessarily be tackled in the order in which they are presented in this book.

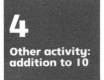

Problem solving

Investigation

Extension

Some pages, or parts of a page, provide an opportunity for problem solving and investigative skills to be applied.

Other pages contain work that may suit the needs of particular children only.

Where a calculator would be useful this is indicated by a calculator symbol.

This symbol indicates that an exercise book is required.

Use tens and units if you wish

This shows that structured materials may be used.

HEINEMANN EDUCATIONAL

Contents

Heinemann Educational,
a division of Heinemann Publishers (Oxford) Ltd,
Halley Court, Jordan Hill, Oxford OX2 8EJ

© Scottish Primary Mathematics Group 1992

First Published 1992

93 94 95   10 9 8 7 6 5 4 3

Designed by Miller, Craig & Cocking
Produced by Oxprint Ltd
Printed in the UK by Scotprint Ltd, Edinburgh.

ISBN 0 435 02094 3

Measure, Shape and Handling data Workbook

		Workbook	Textbook
Time	Revision of hours and half hours	1	
	Concept of a minute	2	
	Minutes past the hour: digital and analogue	3–7	
	Matching digital and analogue times		27
	Quarter past and quarter to		28
	Before and after, simple durations		29
Other activity			30
Length	Metres and half metres	8–10	
	Centimetres	11–14	31
Area	Counting squares	15, 16	32
Volume	Litres	17, 18	
Weight	Kilograms and half kilograms	19, 20	33
Measure	Units, investigation		34, 35
3D Shape	Triangular prism	21	
	Faces, edges and corners	22–24	
Tiling	Using squares, rectangles and regular hexagons	25, 26	
Right angles	Turning through right angles	27, 28	
	Right angles in 2D shapes	29	36
Other activities			37, 38
Symmetry	One line of symmetry	30–32	39, 40
Handling data	Decision diagrams	33, 34	
	Organising and interpreting data, scaled axes	35–38	41–43
	Probability		44
Other activity			45

MYSTERY TOUR

Use the codes to find what was liked on the tour.

1

G	Y	D	A	L
11	12	13	14	15

Copy and complete.

6	8	6	7	5	10	7
+ 5	+ 6	+ 9	+ 7	+ 8	+ 4	+ 5
11	—	—	—	—	—	—
G						

2

A	H	B	E	L	T	W	U	R	D	O	G
4	5	6	7	8	9	10	11	12	13	14	15

Copy and complete.

15	14	13	14	7	8
– 6	– 9	– 6	– 8	+ 4	+ 7
—	—	—	—	—	—

15	8	3	15	6
– 5	+ 6	+ 9	– 7	+ 7
—	—	—	—	—

3 Use the red code.   Write number stories for

R U T H   and   L E E L A

1 Find the car numbers.

- (a) 10 + 10
- (b) 16 – 7
- (c) 8 + 9
- (d) 17 – 10
- (e) 9 + 10
- (f) 18 – 9

2
(a) Find Jan's score.
(b) Find Seb's score.
(c) What is the difference between their scores?

Jan
10 9
Add

Seb
8 9
Add

3 The Green Park roundabout holds 16 children.
There are 9 children on it.
How many more can get on?

4 Find the rule for each machine.

Problem solving Extension

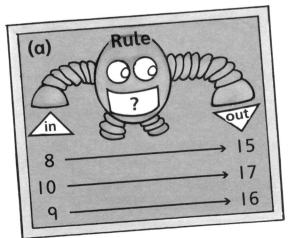

(a) Rule ?
in
8 ——→ 15
10 ——→ 17
9 ——→ 16

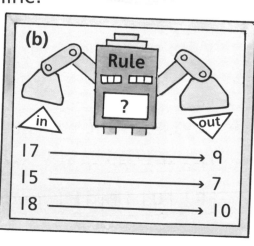

(b) Rule ?
in   out
17 ——→ 9
15 ——→ 7
18 ——→ 10

What we liked best

	Funpark	Bug World	Gala Day	Green Park	Planetarium
8				Seb	
7				Phil	
6				Salma	
5	Irfan			Jan	
4	Kathy			Alison	Sanjit
3	Monica	Karen		Pam	Li
2	Leela	Bill	Carl	Tim	Ruth
1	Wes	Kapil	Sandra	Sam	Ravi

1 Who liked the Gala day best?

2 How many children liked best
(a) Funpark (b) Green Park?

3 How many children altogether went on the tour?

4 The children who liked Green Park best made this diagram.

	Played on the swings	Did not play on the swings
Played on the slide	Seb Alison Sam	Salma Tim
Did not play on the slide	Jan Pam Phil	

Who played on (a) the swings   (b) the slide
                (c) the swings **and** the slide?

5 Find out which of the places on the mystery tour your class likes best. Draw a graph of your results.

Ask your teacher what to do next.

1 **(a)** Play 'Make 10' with a friend.
Each of you copy this number square on
squared paper.

Each loop holds
numbers which
add to 10.

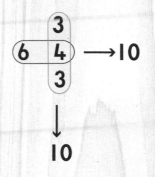

(b) Draw loops round other numbers which add to 10.
Who found more loops?

2 **(a)** Each of you copy this number square.
Now play 'Make 9'.

Make 9	3	4	2	5	4
	5	8	2	1	6
	4	1	5	7	0
	1	7	3	2	3
	4	0	5	1	8

(b) Who found more loops?

Domino fun

You need these four dominoes.

Tracey used the dominoes → and made this number rectangle. →

6	5	4	3
5	4	3	6

1 Use the dominoes a different way to make the same number rectangle.

2 Now make each of these number rectangles.

(a)

4	5	4	5
3	6	3	6

(b)

6	5	6	5
3	4	3	4

(c)

5	4	3	3
4	5	6	6

(d)

3	6	6	4
4	5	5	3

3 Draw a number rectangle for these four dominoes.

Ask a friend to use the dominoes to make your number rectangle.

Find the answer for each step.
Write the word which tells you to add.

sum

plus

total

1 What is the total score?

5 6 8

2 Find the sum of 39 and 58.

3 How much is 50p plus 35p?

4 Find each total:

(a)	(b)	(c)	(d)
4 8	3 9	6 2	4 0
+3 1	+5 4	+1 8	+2 5

5 Find the sum of:
(a) 42 and 30 (b) 40 and 53
(c) 35 and 23 (d) 24 and 46.

6 Make up a question about adding.
Find the answer.

Problem solving

8
(a) Find two numbers which total 64.
(b) Find two **different** numbers which total 64.

9
Is this a magic square?

30	12	17
18	14	27
11	33	15

10
Find **three** numbers whose sum is 66.

7
Find two numbers which add to 48.

6
(a) Find the total for 1 + 2 + 3 + 4 + 5.
(b) Guess the total for 11 + 12 + 13 + 14 + 15. Now check.

Go along the problem path.

Use a calculator.

1
(a) Find 14 + 14 + 14 + 14 + 14.
(b) Guess the answer to 15 + 15 + 15 + 15 + 15.

11

Use each of the keys
3, **4** and **5** once.
Make this addition
? **?** **+** **?** **=**
have the greatest
total.

12

Use the keys
2, **4**, **5**, **+** and **=**
as often as you like to
make ⟨ 2 1. ⟩

2 1.

5

(a) Which is the greatest
    number 32, 28 or 36?

(b) Add 24 to each.
    Which is the greatest
    answer?

4

Is this a
magic
square?

18	14	34
38	22	6
10	30	26

2

(a) Find 16 + 16 + 16 + 16 + 16.

(b) Guess the answer to
    17 + 17 + 17 + 17 + 17.

3

Guess how many
times you press
1 **8** **+**
to make ⟨ 90. ⟩

90.

Ask your teacher what to do next.

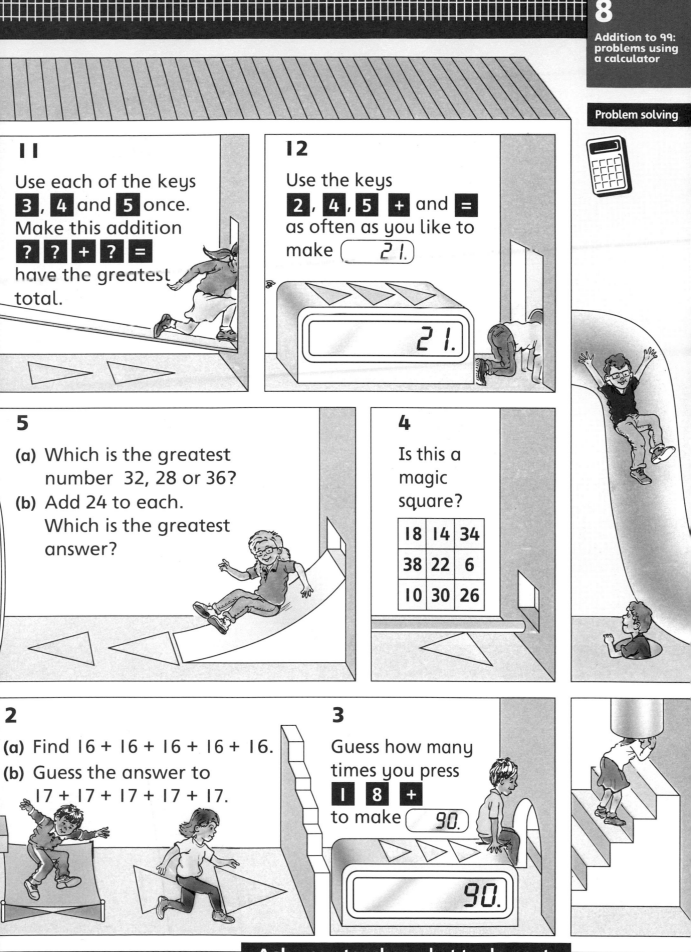

Trail Take Away

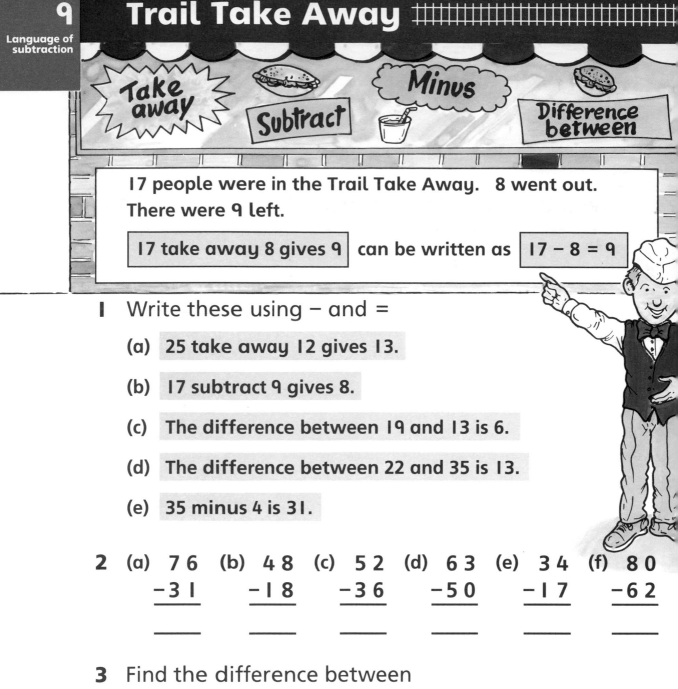

Take away · Subtract · Minus · Difference between

17 people were in the Trail Take Away. 8 went out.
There were 9 left.

| 17 take away 8 gives 9 | can be written as | 17 − 8 = 9 |

1 Write these using − and =

(a) 25 take away 12 gives 13.

(b) 17 subtract 9 gives 8.

(c) The difference between 19 and 13 is 6.

(d) The difference between 22 and 35 is 13.

(e) 35 minus 4 is 31.

2 (a) 7 6   (b) 4 8   (c) 5 2   (d) 6 3   (e) 3 4   (f) 8 0
       −3 1   −1 8   −3 6   −5 0   −1 7   −6 2
      ────   ────   ────   ────   ────   ────

3 Find the difference between
(a) 61 and 28   (b) 23 and 50.

4 (a) 97 children went to the Trail Take Away in the
    morning. 78 went in the afternoon.
    What is the difference in the number of children?

(b) The cook fried 30 fish. He had orders for 42 fish.
    How many more did he need to fry?

5 Write a story about the Trail Take Away
    for 19 − 6 = 13.

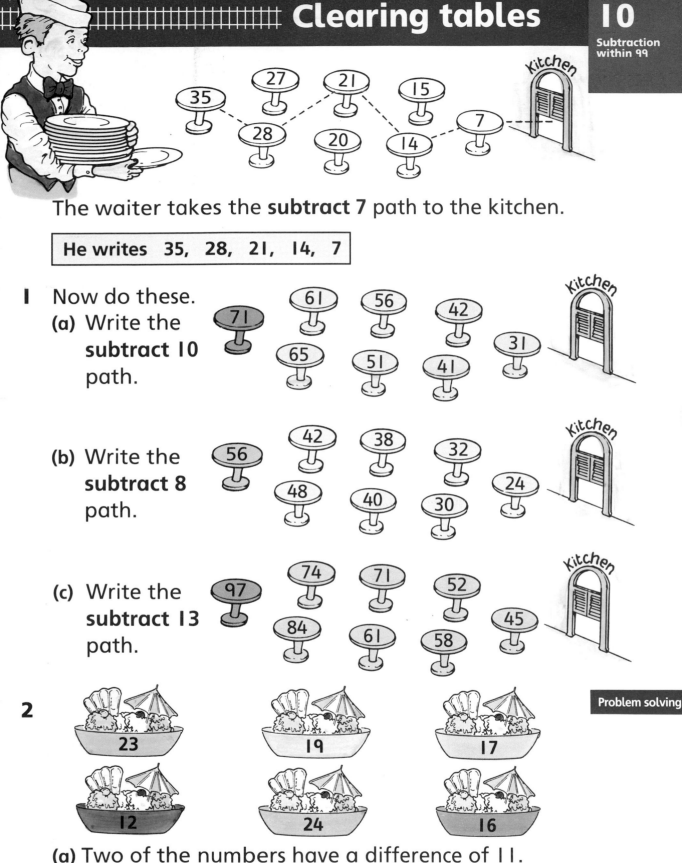

The waiter takes the **subtract 7** path to the kitchen.

| He writes   35,   28,   21,   14,   7 |

1 Now do these.

(a) Write the **subtract 10** path.

(b) Write the **subtract 8** path.

(c) Write the **subtract 13** path.

2

Problem solving

(a) Two of the numbers have a difference of 11.
   Write these numbers.
(b) Write a pair of numbers with a difference of 7.
(c) Find other pairs with a difference of 7.

1 (a)

17 − 8 =

27 − 8 =

37 − 8 =

47 − 8 =

Guess these answers and then check.

57 − 8 =

67 − 8 =

77 − 8 =

87 − 8 =

(b)

Find these answers.

16 − 9 =

26 − 9 =

36 − 9 =

Continue the pattern.

=

=

=

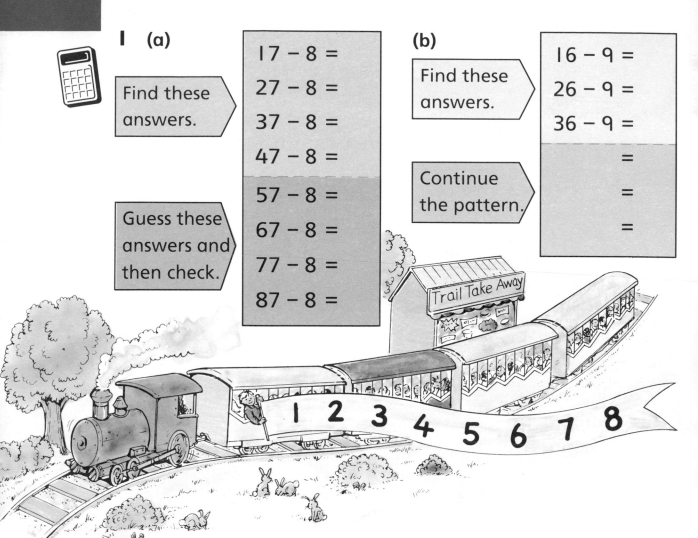

Trail Take Away

1 2 3 4 5 6 7 8

2 (a) Find the difference between
  • the first and second numbers
  • the second and third numbers.

| 12 | 34 | 56 | 78 |

(b) Guess the difference between the third and fourth numbers.   Check.

3 Do the same for these four numbers.
What do you notice?

| 87 | 65 | 43 | 21 |

4 Do the same for these four numbers.
What do you notice?

| 18 | 27 | 36 | 45 |

1 Look at this table.

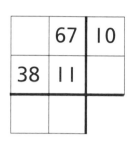

Subtract

14	7 → 7
12	2 → 10
↓ 2	↓ 5

Now copy and complete these tables.
Use squared paper.

18	9	
14	6	

36	14	
22		12

	67	10
38	11	

2 Copy and put the correct number in each ☐ and △.

(a)
```
   ☐ 8
 − 3 △
 ───────
   2 2
```

(b)
```
   ☐ 6
 − 3 4
 ───────
   2 △
```

(c)
```
   5 △
 − ☐ 9
 ───────
   2 3
```

3

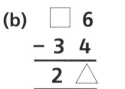

30
44   62
27   14
49   31
17

Each number has a partner.

(a) Find the difference
    between each number
    and its partner.

(b) Find a partner for each
    of these:
    60 and 86 .

(c) Find another partner
    for each.

Ask your teacher what to do next.

26 boys and 17 girls on board

16 boys and 19 girls on board

1 Which boat has more boys? How many more?

2 Find the total number of children in **(a)** the red boat
                                                        (b) the blue boat

3 Which boat has more children? How many more?

4 How many children are there altogether in the two boats?

5 Look at the seat numbers of the **back row**.

39 40   41 42

35 36   37 38

31 32   33 34

(a) Add the brown seat numbers, 39 and 42.
(b) Add the green seat numbers, 40 and 41.
(c) What do you notice about your answers?

6 Do the same again for the seat numbers of the other two rows.

1 Each log has 10 seats.
How many seats
are there altogether?

2 The last log
has 7 empty seats.
The other logs are full.
How many people are
in the logs altogether?

3 Fourteen people are adults.
How many are children?

4 Look at the log seat numbers.

(a) Add the seat numbers which face each other,
30 + 21, 29 + 22 and so on.

(b) What do you notice about your answers?

5 (a) Find the difference between the seat numbers
which face each other, 30 − 21, 29 − 22 and so on.

(b) What do you notice about your answers
this time?

Ask your teacher what to do next.

1 Make 4 cards like these.   **2**   **3**   **4**   **5**

2 (a) Use the cards to make
two numbers which add to **77**.

(b) Use the cards again to
find another way of
making **77**.

+

7 7

3 Use the cards to make a total of **(a)** 59   **(b)** 68.

4 Change the   **2**   card for a   **6**   card.

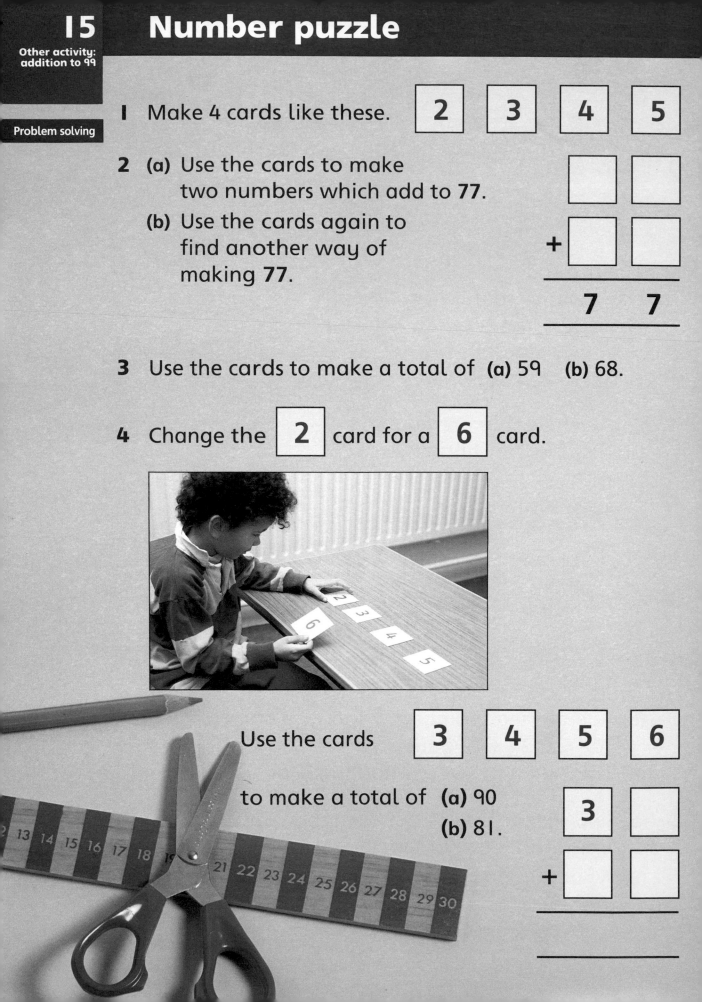

Use the cards   **3**   **4**   **5**   **6**

to make a total of **(a)** 90
(b) 81.

3

+

Other activity:
complementary
addition

1 Guess the missing number.

17 + ? = 24

Use your calculator to check like this:

17.　+ 7 =　24.

2 Guess each missing number and write it down.
Use your calculator to check.
Tick if your guess is correct.

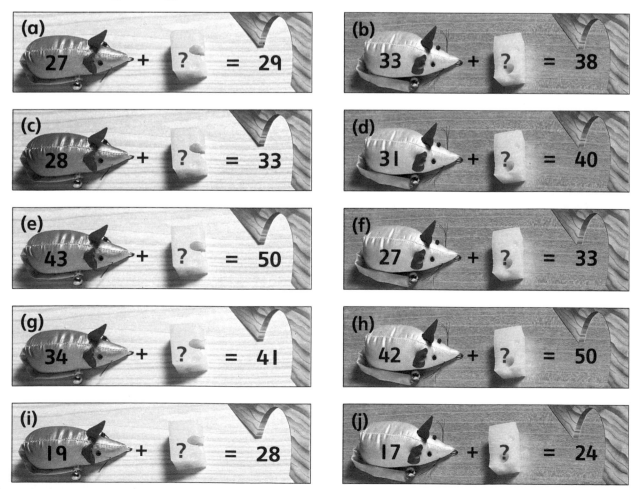

(a) 27 + ? = 29

(b) 33 + ? = 38

(c) 28 + ? = 33

(d) 31 + ? = 40

(e) 43 + ? = 50

(f) 27 + ? = 33

(g) 34 + ? = 41

(h) 42 + ? = 50

(i) 19 + ? = 28

(j) 17 + ? = 24

Paint trays

I How many paints are there on
   (a) 2 red trays   (b) 3 blue trays   (c) 2 green trays
   (d) 3 yellow trays (e) 3 red trays   (f) 2 blue trays
   (g) 3 green and I red tray (h) 3 yellow and I blue tray
   (i) 3 red and 2 green trays
   (j) 2 blue and 3 yellow trays?

2 (a) How many brushes are
       there in 2 packets?

   (b) How many crayons are
       there in 3 boxes?

   (c) How many pencils are
       there in 3 boxes?

   (d) How many paint blocks
       are there in 2 packs?

   (e) How many paint tubs
       are there in 3 boxes?

I Find the cost of

(a) 2 crayons   (b) 3 tubes of paint   (c) 2 paint blocks

(d) 3 pencils   (e) 2 chalks   (f) 3 crayons

(g) 3 chalks and I pencil   (h) 2 crayons and I paint block

(i) 2 tubes of paint and 3 paint blocks.

2 Tom and Jane each had 20 pence.

(a) Tom bought 3 chalks.
What was his change?

(b) Jane bought 2 pencils.
What was her change?

3 (a) Eric bought 2 of the **same** item.
He spent 20p.   What did he buy?

Problem solving

(b) Abda bought 3 of the same item.
She spent 24p.   What did she buy?

4 Freda bought a red pencil and green chalk.

(a) How much did she spend?

(b) She paid with 3 coins of the same value.
What value was each coin?

Ask your teacher what to do next.

At home

1 How many of each animal has Ling in her book? She has

(a) 2 pages with 9 lions on each

(b) 3 pages with 7 tigers

(c) 4 pages with 6 bears

(d) 5 pages with 4 elephants

(e) 10 pages with 3 zebras.

2 How many players of each sport has Jamal in his book? He has

(a) 10 pages with 5 skaters on each page

(b) 3 pages with 6 swimmers

(c) 2 pages with 7 footballers

(d) 4 pages with 8 golfers

(e) 5 pages with 9 runners.

3 The table shows how many stickers each child bought. How much did each child spend altogether?

	5p	8p	9p	6p	7p
Ling		4		3	
Jamal			5		2
Stefan	3	10			
Sita	1		4		5

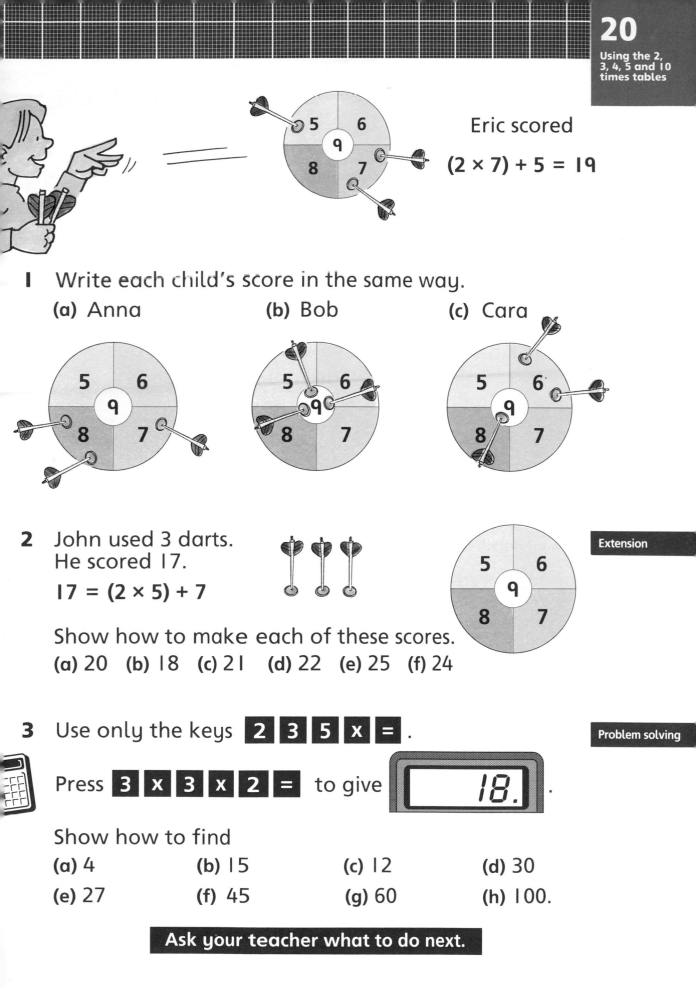

Eric scored

$(2 \times 7) + 5 = 19$

1 Write each child's score in the same way.

(a) Anna (b) Bob (c) Cara

2 John used 3 darts.
He scored 17.

$17 = (2 \times 5) + 7$

Extension

Show how to make each of these scores.
(a) 20 (b) 18 (c) 21 (d) 22 (e) 25 (f) 24

3 Use only the keys **2** **3** **5** **x** **=** .

Problem solving

Press **3** **x** **3** **x** **2** **=** to give ⌐ *18.* ⌐ .

Show how to find
(a) 4   (b) 15   (c) 12   (d) 30
(e) 27   (f) 45   (g) 60   (h) 100.

Ask your teacher what to do next.

1 Kim put out twelve cakes in **2 rows of 6.**

Use counters.

(a) Find and draw other ways of putting out
12 cakes with an equal number in each row.

(b) Do the same for 20 cakes.

2 Dad sits opposite Mum .

Tracey sits on Dad's right.

Jason sits opposite Tracey.

Kim sits between Mum and Tracey.

Where does Scott sit?

Mum

A game for 4 players.

Each player needs 10 cubes, a counter and a track.
Ask your teacher how to play.

Start here ▼

2

3

Put 4 cubes in the remainder box.

4

4

4

2

Put 4 cubes in the remainder box.

2

3

3

Put 3 cubes in the remainder box.

4

3

3

Put 3 cubes in the remainder box.

4

4

2

Put 3 cubes in the remainder box.

Ask your teacher what to do next.

Start with 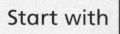 **0.** each time.

1 Add on 1 by pressing **+** **1**. Press **=** seven times.
   (a) What number is shown?
   (b) What happened each time you pressed **=** ?

2 Add on 2 by pressing **+** **2** .
   (a) How many times do you
   think you need to press **=** to make  **8.** ?
   (b) Now check.

3 How many times do you think you need to press **=**
   to make the target?
   Check each time.

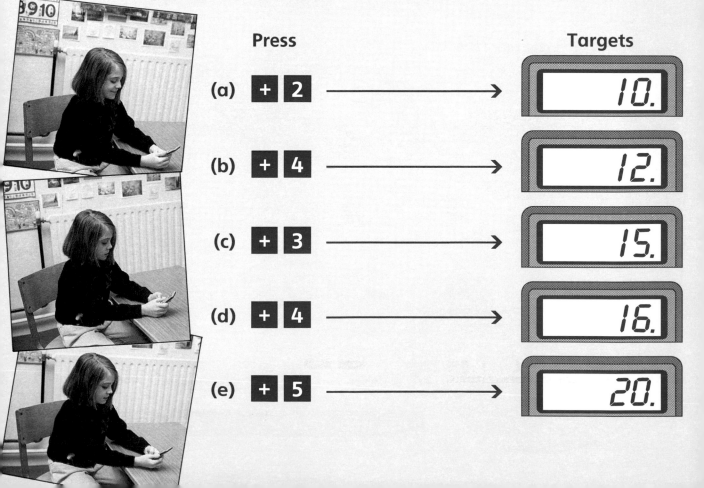

Press		Targets
(a) **+** **2** ⟶		**10.**
(b) **+** **4** ⟶		**12.**
(c) **+** **3** ⟶		**15.**
(d) **+** **4** ⟶		**16.**
(e) **+** **5** ⟶		**20.**

1 Start with **5.**

Take away 1 by pressing **–** **1**. Press **=** five times.
(a) What number is shown?
(b) What happened each time you pressed **=** ?

2 Start with **10.**

Take away 2 by pressing **–** **2**.
(a) How many times do you think you need to press **=** to make **0.** ?
(b) Now check.

3 How many times do you think you need to press **=** to make **0.** ?
Check each time.

Start with	Press
(a) 12.	**–** 4
(b) 20.	**–** 5
(c) 18.	**–** 2
(d) 21.	**–** 3

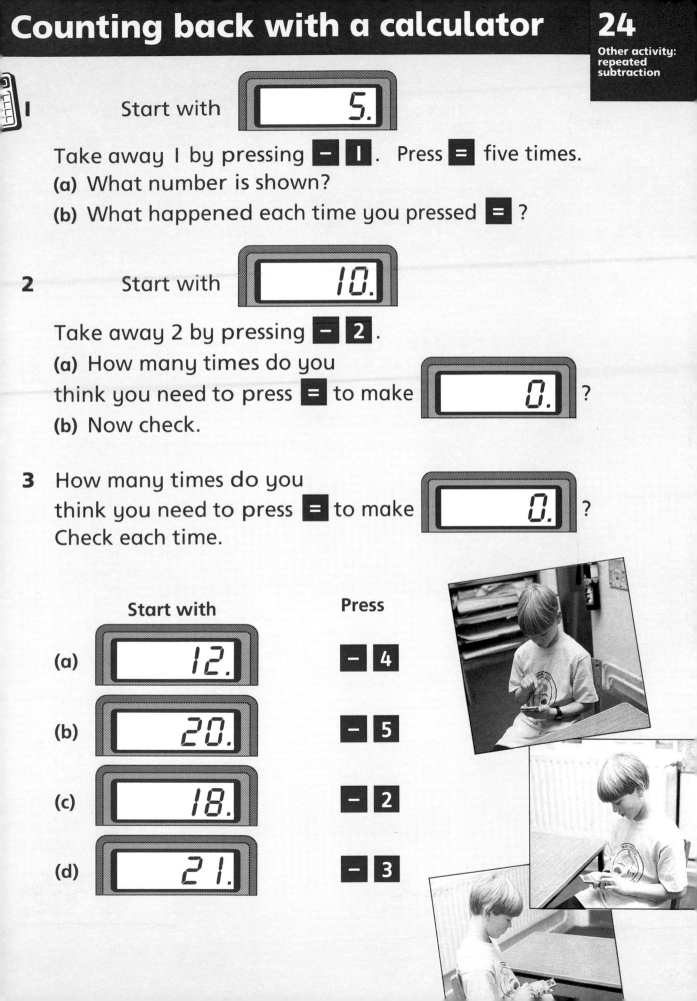

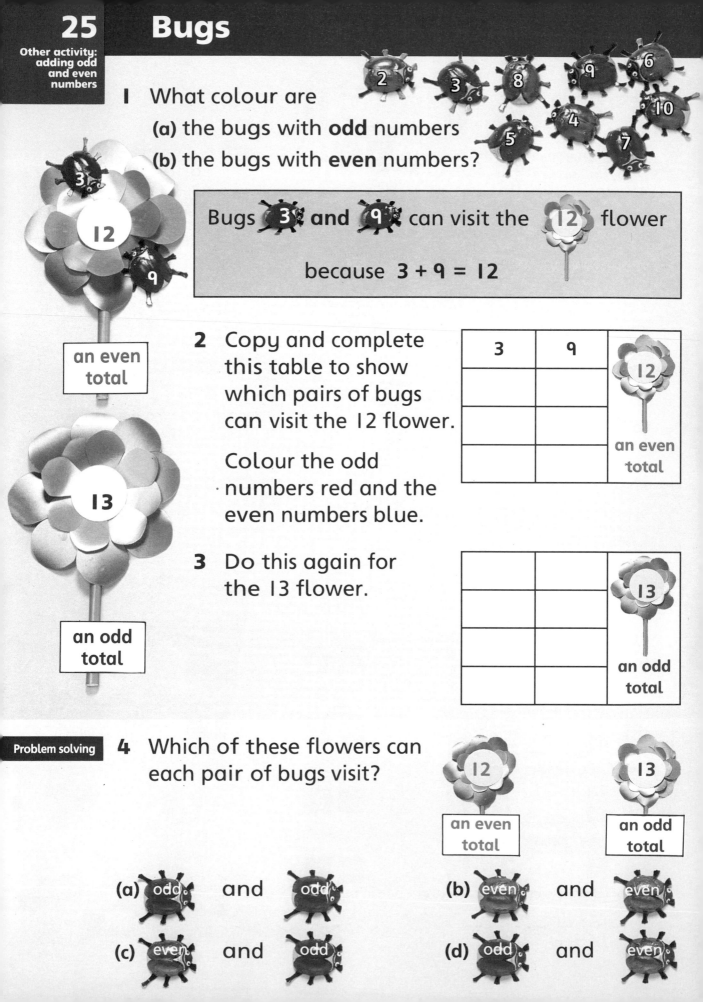

1 What colour are

(a) the bugs with **odd** numbers

(b) the bugs with **even** numbers?

Bugs **3** and **9** can visit the **12** flower

because **3 + 9 = 12**

12

an even
total

13

an odd
total

2 Copy and complete
this table to show
which pairs of bugs
can visit the 12 flower.

Colour the odd
numbers red and the
even numbers blue.

3	9	12
		an even total

3 Do this again for
the 13 flower.

		13
		an odd total

Problem solving

4 Which of these flowers can
each pair of bugs visit?

12
an even
total

13
an odd
total

(a) odd and odd

(b) even and even

(c) even and odd

(d) odd and even

These bees move **up** number trails.

They move like this 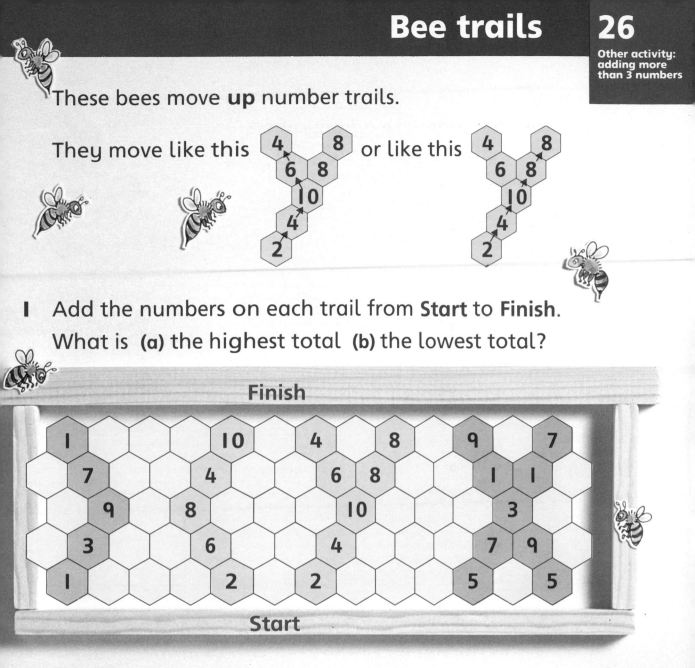 or like this

1 Add the numbers on each trail from **Start** to **Finish**.

What is **(a)** the highest total **(b)** the lowest total?

Finish

Start

2 Find and write **odd** numbers to complete the red trail. It must have a total of between 25 and 35.

3 Find **even** numbers to complete the blue trail. It must have a total of between 24 and 30.

Extension
Problem solving

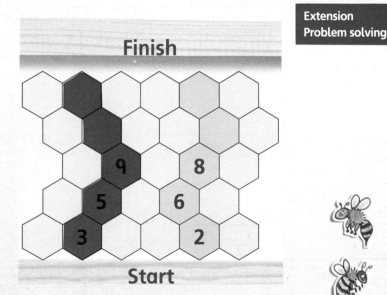

Finish

Start

Clocks and watches

These clocks show
the same time

2:40

1 Write the digital times for each of these.

(a)

(b)

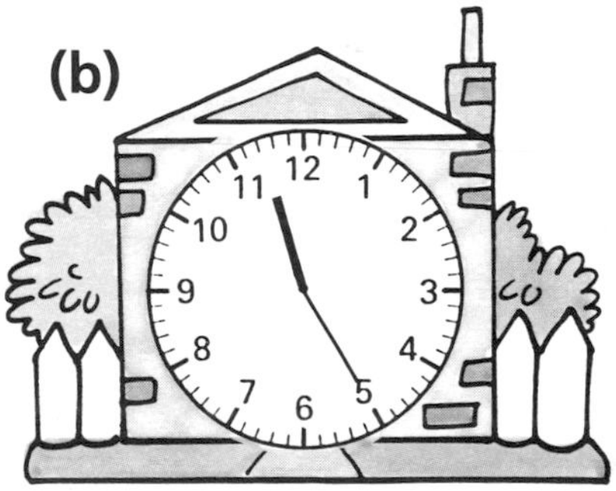

(c)

(d)

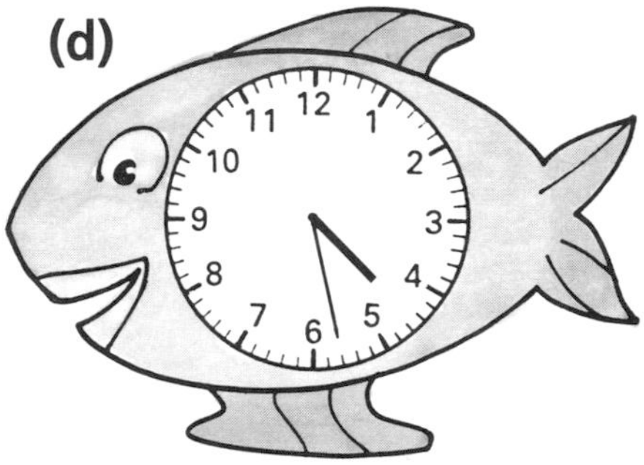

(e)

(f)

2 Write the names of the children whose watches
show the same time.

John

6:45 Moira

Zoe

Tom

6:58

Asif

7:09

Mary

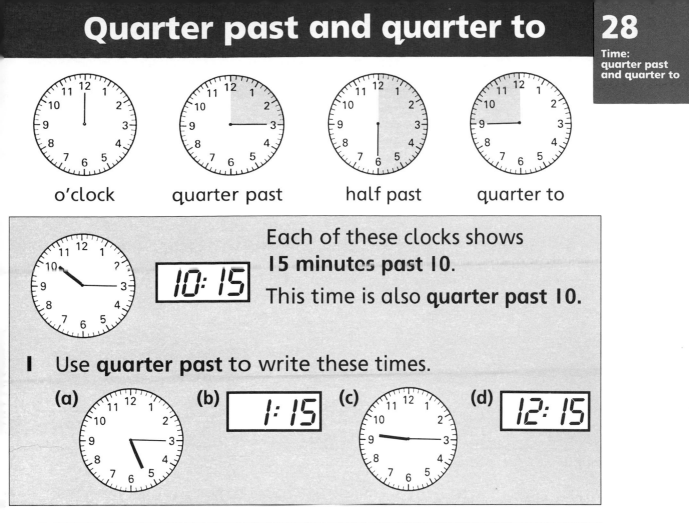

o'clock   quarter past   half past   quarter to

Each of these clocks shows
15 minutes past 10.
This time is also **quarter past 10.**

10:15

1   Use **quarter past** to write these times.

(a)   (b) 1:15   (c)   (d) 12:15

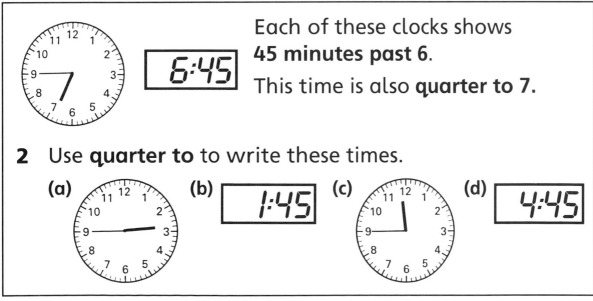

Each of these clocks shows
45 minutes past 6.
This time is also **quarter to 7.**

6:45

2   Use **quarter to** to write these times.

(a)   (b) 1:45   (c)   (d) 4:45

3   Write these times.

(a) 8:45   (b)   (c) 11:15   (d)

Before and after

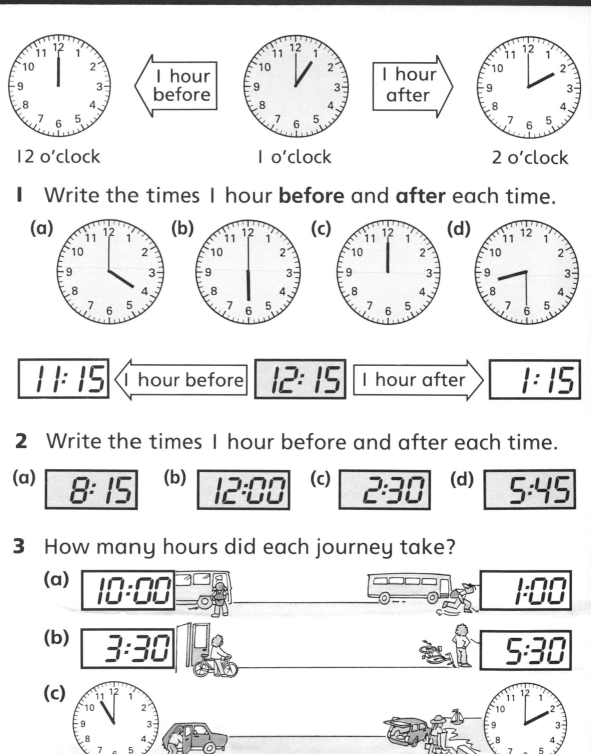

12 o'clock   1 o'clock   2 o'clock

1 Write the times 1 hour **before** and **after** each time.

(a)   (b)   (c)   (d)

| 11:15 | 1 hour before | 12:15 | 1 hour after | 1:15 |

2 Write the times 1 hour before and after each time.

(a) 8:15   (b) 12:00   (c) 2:30   (d) 5:45

3 How many hours did each journey take?

(a) 10:00   1:00

(b) 3:30   5:30

(c)

(d)

Ask your teacher what to do next.

JANUARY

Sun	Mon	Tue	Wed	Thu	Fri	Sat
		1	2	3	4	5
6	7	8	9	10	11	12
13	14	15	16	17	18	19
20	21	22	23	24	25	26
27	28	29	30	31		

1 What month is shown on this calendar?

2 On which day does the month **(a)** start **(b)** finish?

3 How many days are in the month?

4 What day of the week is
(a) 3 January **(b)** 14 January **(c)** 25 January?

You need a calendar for **this year**.

Investigation

5 Write the names of the months with **(a)** 31 days **(b)** 30 days.

6 **(a)** What month is missing? **(b)** How many days has it?

Work with a friend.

1 Use a tape measure to find
   (a) how wide the top of the waistcoat should be
   (b) how long the waistcoat should be.

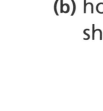
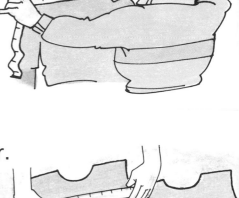

2 You need large sheets of paper. Measure and cut out two shapes like these.

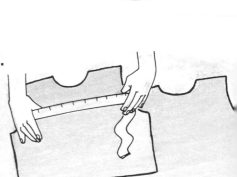

3 Cut one shape down the middle to make the front. Join the front and back pieces with tape.

4 Use coloured shapes to make a pattern on the waistcoat.

Ask your teacher what to do next.

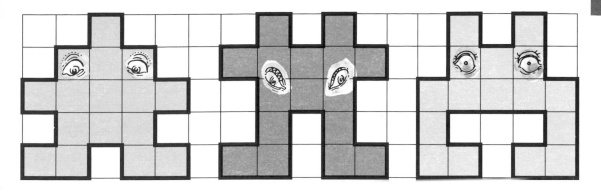

1 (a) Which alien do you **think** has the largest area?

(b) Check by finding the area of each alien in squares.

(c) Write about what you found out.

2 Here are some other aliens. Only **half** of each is shown. What will be the area of each **whole** alien?

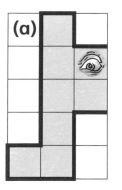

(a)

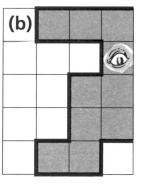

(b)

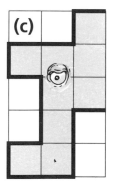

(c)

3 You need squared paper. Draw your own alien with an area of 24 squares.

Ask your teacher what to do next.

The tin of fruit weighs more than $\frac{1}{2}$ kilogram.

The melon weighs less than 1 kilogram.

1 Write a story about the weight of each object below.

(a) cabbage — 1 kg

(b) tin — $\frac{1}{2}$ kg

(c) butter — $\frac{1}{2}$ kg

(d) 2 kg — marrow

(e) $\frac{1}{2}$ kg — carrot

(f) $1\frac{1}{2}$ kg — turnip

2 Write the weight of each object.

(a) CORN FLAKES — $\frac{1}{2}$ kg

(b) Flour — $\frac{1}{2}$ kg, 1 kg

(c) 1 kg, $\frac{1}{2}$ kg, 1 kg — potatoes

Ask your teacher what to do next.

Choose and write
the correct label for each.

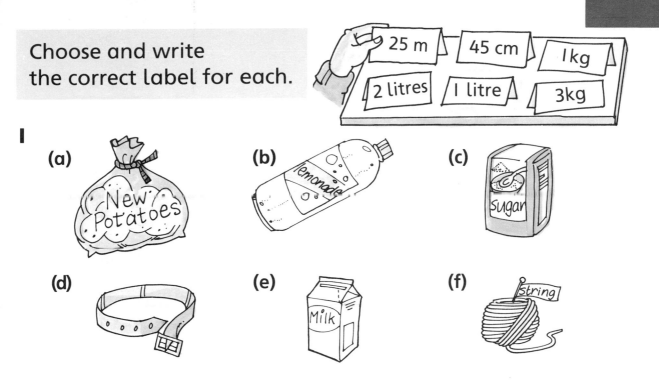

25 m   45 cm   1 kg

2 litres   1 litre   3 kg

1

(a)   (b)   (c)

(d)   (e)   (f)

Choose the best measure for each.

2
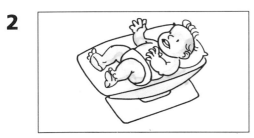
A new born baby weighs about
(a) $\frac{1}{2}$ kg   (b) 15 kg   (c) 3 kg.

3
A kettle holds about
(a) 2 litres   (b) 5 litres   (c) 7 litres.

4

The height of the classroom is about
(a) 10 m   (b) 2 m   (c) $3\frac{1}{2}$ m.

1 A giraffe is about 5 metres tall.

Would its head touch the ceiling of your classroom?

2 A dog drinks about 1 litre of water a day.

About how many saucersful is this?

3 About how many apples are there in $\frac{1}{2}$ kg?

4 A caterpillar crawled 3 metres in a minute.

Measure out 3 metres. Take 1 minute to walk this distance.

Ask your teacher what to do next.

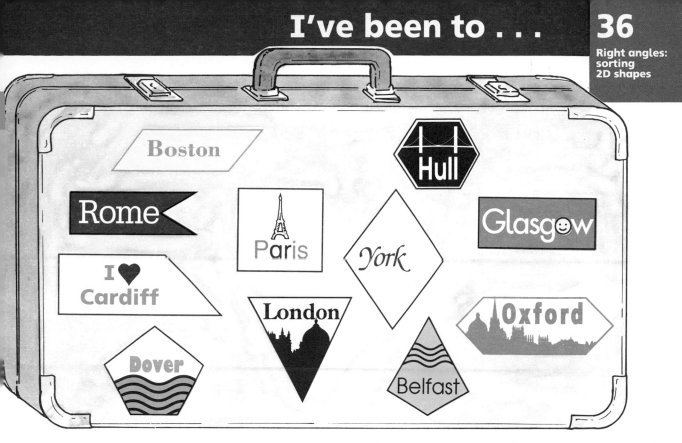

	Right angles	~~Right angles~~
4 sides	Cardiff	
4 ~~sides~~		

1 Draw a large Carroll diagram. Write the place names.

2 Take each shape through the tree diagram.
Write the place names for (a), (b), (c) and (d).

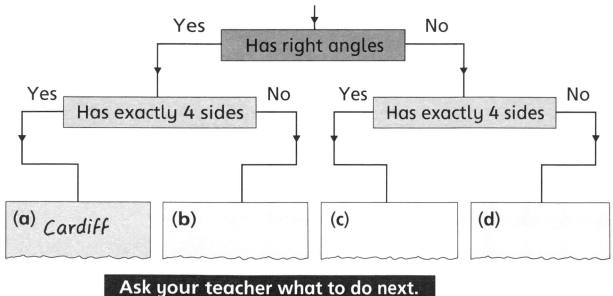

Ask your teacher what to do next.

Problem solving | **You need a set of dominoes.**

1 Use **four** dominoes to make each shape.

(a) **(b)** **(c)** **(d)**

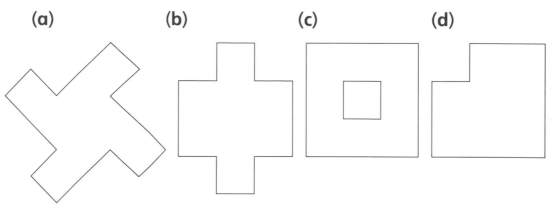

2 Use **six** dominoes to make a train like this.

3 In this train there is a difference of 1 **at each join.**

join   join

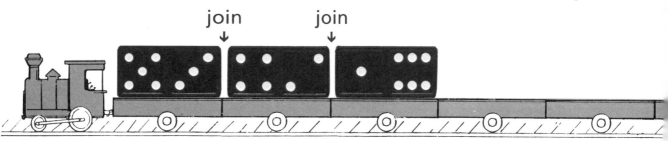

Use **six** dominoes to make a train like it.

This domino has a difference of 1.  5 − 4 = 1

4 How many dominoes have
    (a) a difference of 1   **(b)** a difference of 2
    (c) a difference of 3   **(d)** a difference of 4?

1 Can you see the five squares
in this drawing?

How many squares can you see in each drawing?

(a)   (b)   (c)

2 **Use squared paper.**
(a) Cut out 4 shapes like this ⟶
(b) Place the 4 shapes together
to make a **square**.
(c) Stick the square in your exercise book.

3 Do the same again for
4 shapes like this ⟶

4 Do the same again for
4 shapes like this ⟶

T-shirts

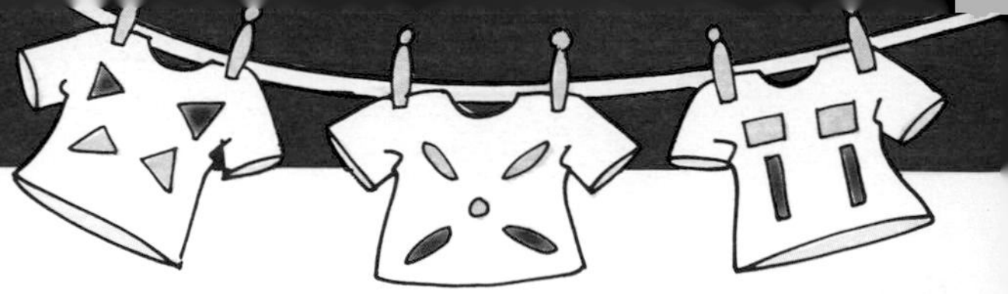

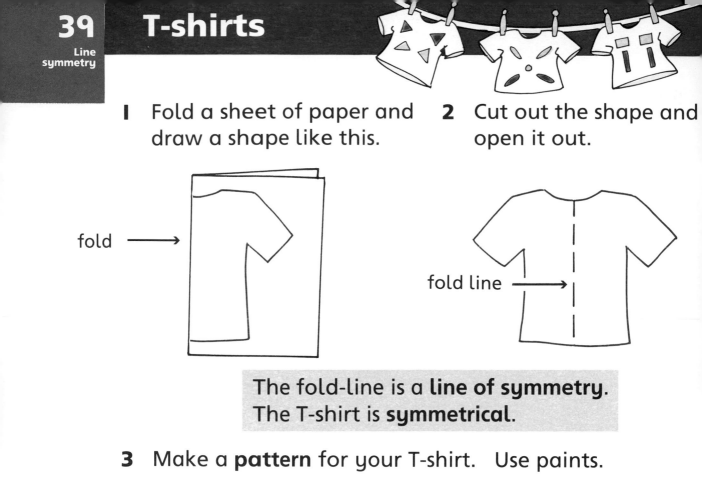

1 Fold a sheet of paper and draw a shape like this.

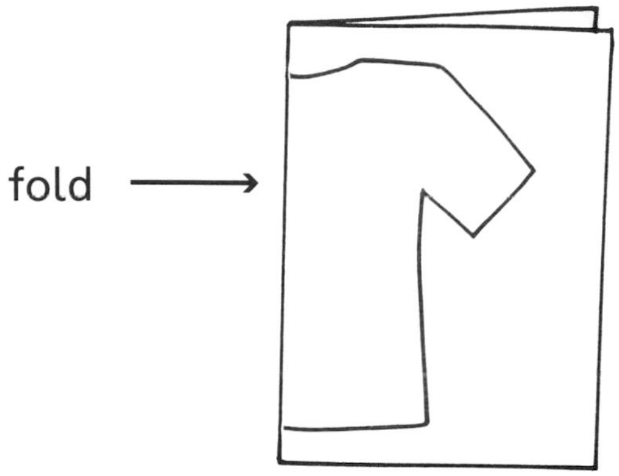

2 Cut out the shape and open it out.

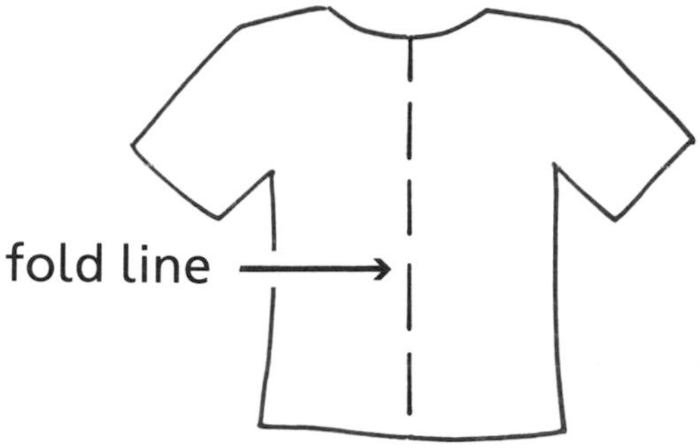

The fold-line is a **line of symmetry**.
The T-shirt is **symmetrical**.

3 Make a **pattern** for your T-shirt. Use paints.

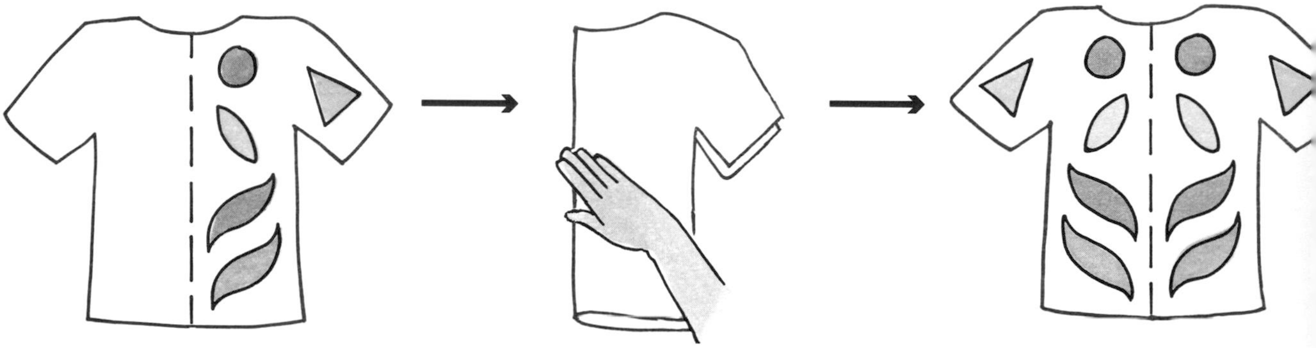

Your T-shirt **pattern** is symmetrical.

Problem solving

4 Make symmetrical shorts.

Mark the line of symmetry.

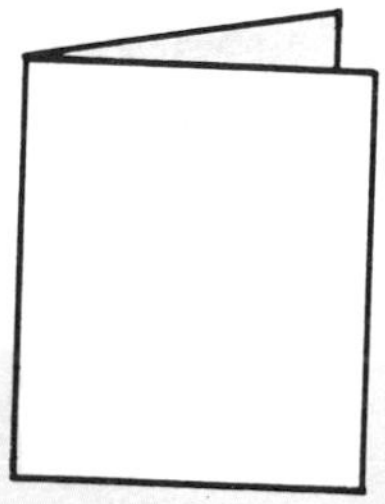

Go back to Measure, Shape and Handling data Workbook page 30, question 1.

Place a mirror on each dotted line.

1 Write the name of each whole picture.

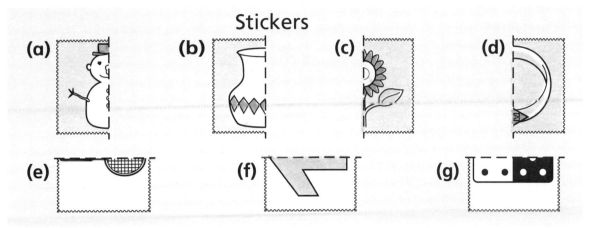

Stickers

2 Name the shape of each whole key-ring.
(a) is a rectangle.

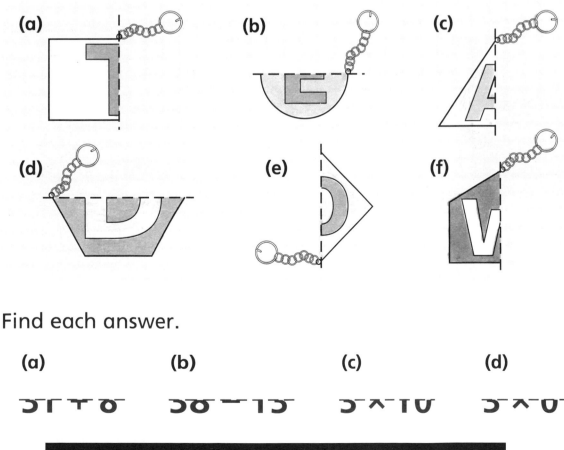

3 Find each answer.

(a)   (b)   (c)   (d)

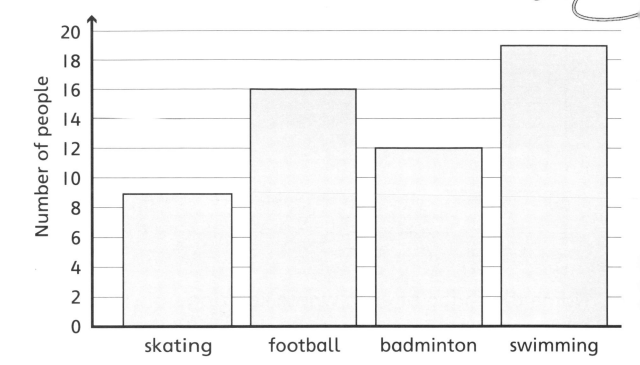

What people did at the Centre

1 Which of these sports did
(a) most people play   (b) fewest people play?

2 How many people played each of the sports?

3 The Leisure Centre wants to offer another sport.

skate boarding   roller skating   cycling

(a) Make a tick sheet. Ask the children in your class
which of these sports it should be:

skate boarding	
roller skating	
cycling	

(b) Draw a graph to show your results.
(c) Write three sentences about your graph.

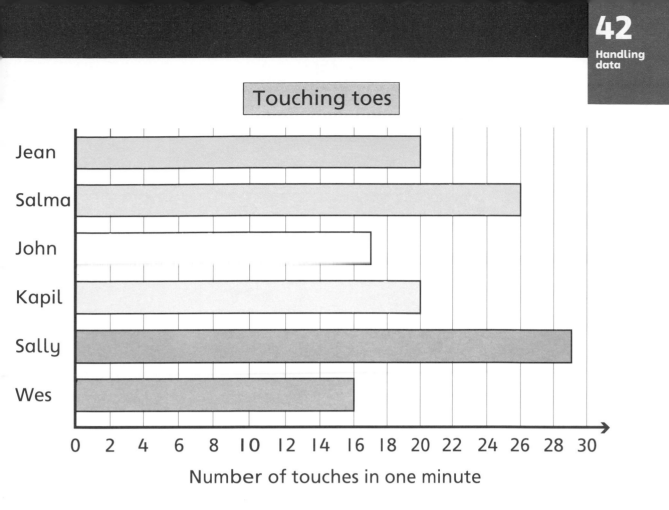

Touching toes

Number of touches in one minute

1 (a) Who touched their toes most often?

(b) Which two children had the same score?

(c) What did each child score?

(d) What is the difference between
  • Jean's and Salma's score?   • Sally's and Wes's scores?

2 Your group should choose one of
these activities.

stand up – sit down   bouncing ball

step-ups on a bench

(a) Each of you do it for one minute.

(b) Make a list of the scores.

(c) Draw a graph of the scores.

(d) Write three sentences about your graph.

1 Name the film which **(a)** costs most **(b)** costs least.

2 Which film starts at **(a)** 7 o'clock **(b)** half past seven?

3 Which film finishes at **(a)** 8:45 **(b)** quarter past 9?

4 Which films are not shown on Thursday?

Extension

5 You are watching a film in the cinema at half past nine. Which film are you watching?

6 Find out about films being shown at a cinema near you.

For each event write
very likely, likely, unlikely or **very unlikely**.

1 Someone will drop litter in our playground today.

2 All the sweets in the smartie tube will be yellow.

3 The cinema will open at 7 o'clock in the morning.

4 I will use a calculator today.

5 A parent will visit our class today.

6 Our teacher will tell us a story today.

Ask your teacher what to do next.

Maths Bar Patterns

1 Use squared paper. Copy and complete each **pattern**

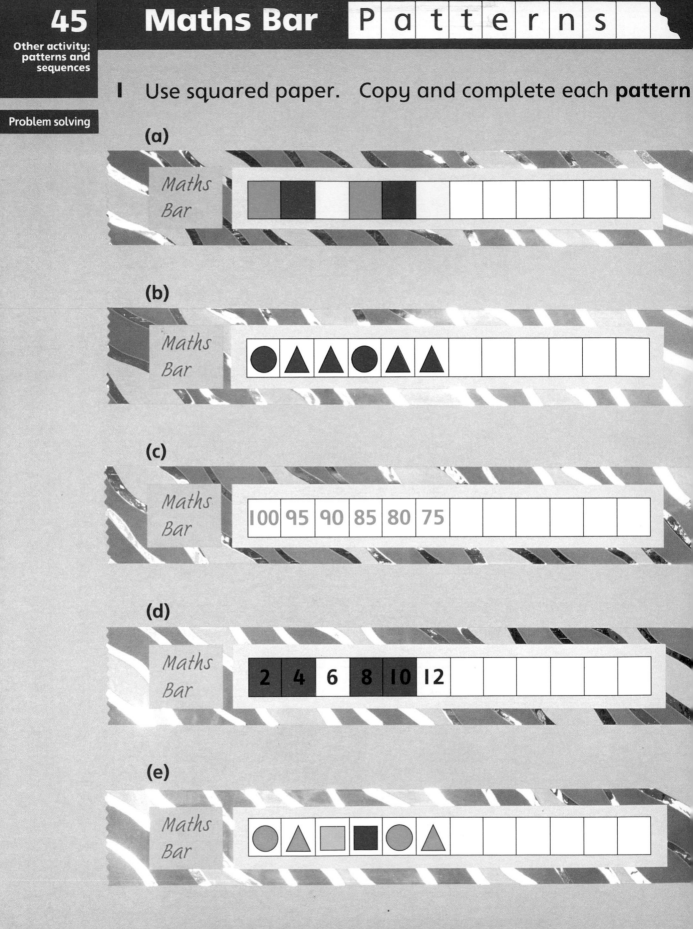

(a)

Maths Bar

(b)

Maths Bar

(c)

Maths Bar

| 100 | 95 | 90 | 85 | 80 | 75 | | | | | | |

(d)

Maths Bar

| 2 | 4 | 6 | 8 | 10 | 12 | | | | | | |

(e)

Maths Bar

2 Make your own *Maths Bar* pattern.